Pharmaceutical Microbiology Glossary

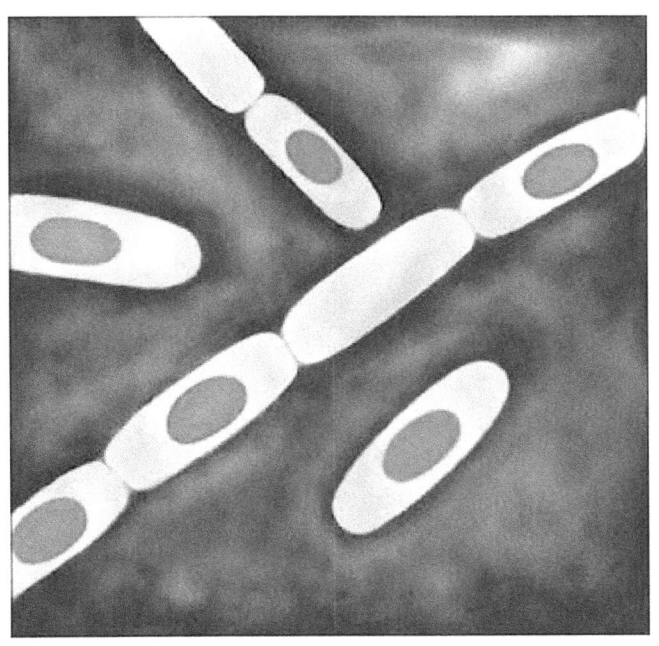

2nd Edition

By Dr. Tim Sandle

Tim Sandle

Produced by Microbiology Solutions.

First Edition: Published August 2012
Second Edition: Published September 2013

Website: http://www.pharmamicroresources.com/

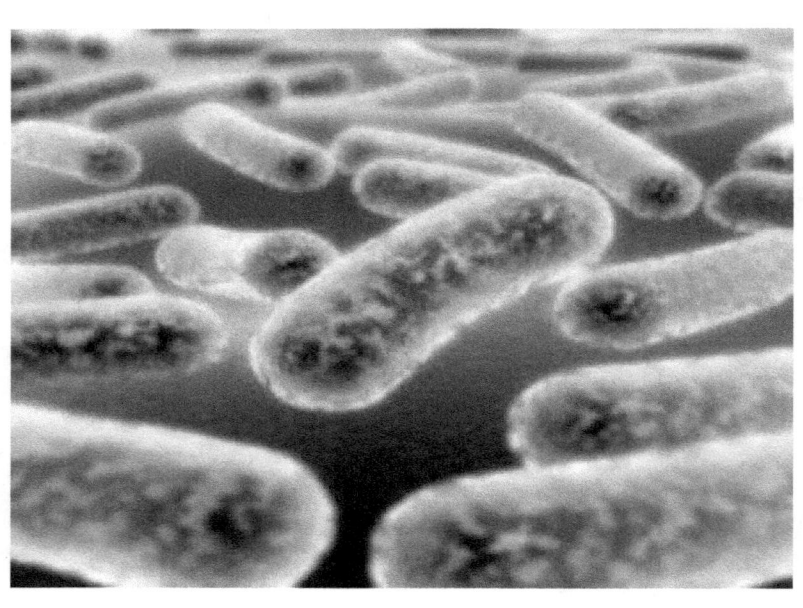

About the author

Tim Sandle, Ph.D, M.A., BSc (Hons), CBiol, MSBiol., MIScT

Dr. Sandle is a chartered biologist (Society for Biology) and holds a first class honors degree in Applied Biology; a Masters degree in education; and obtained his doctorate from Keele University.

Dr. Sandle has over twenty years experience as a pharmaceutical microbiologist, which includes designing, operating and reviewing a range of microbiological tests (including sterility testing, endotoxin LAL methodology, microbial enumeration, environmental monitoring, particle counting and water testing). In addition, Dr. Sandle is experienced in microbiological and quality batch review, microbiological investigation and policy development.

In addition, Dr. Sandle is an honorary consultant with the School of Pharmacy and Pharmaceutical Sciences, University of Manchester and is a tutor for the university's pharmaceutical microbiology M.Sc course. Dr. Sandle serves on several national and international committees relating to pharmaceutical microbiology and cleanroom contamination control (including the ISO cleanroom standards). He is currently chairman of the Pharmaceutical Microbiology Interest Group (Pharmig) LAL action group and serves on the National Blood Service cleaning and disinfection committee. He has written over one hundred book chapters, peer reviewed papers and technical articles relating to microbiology. Dr. Sandle has also delivered papers to over thirty conferences.

Dr. Sandle is the editor of the Pharmaceutical Microbiology Interest Group Journal and runs an on-line microbiology website and forum (www.pharmig.blogspot.com). Dr. Sandle is an experienced auditor and frequently acts as a consultant to the pharmaceutical and healthcare sectors.

Other books by Tim Sandle include:

Saghee, M.R., Sandle, T. and Tidswell, E.C. (Eds.) (2011): Microbiology and Sterility Assurance in Pharmaceuticals and Medical Devices, New Delhi : Business Horizons

Sandle, T. (2011): Two extremes? Flexible working in Europe: A study of differences in Flexible Working Time between two European Plasma Fractionators in Britain and the Netherlands, Saarbrucken: VDM Publishing, IBSN 978-3-639-34965-8

Sandle, T. (2012). The CDC Handbook: A Guide to Cleaning and Disinfecting Cleanrooms, Grosvenor House Publishing: Surrey, UK

Sandle, T. (2012). E-Guide to Cleanrooms, Microbiology Solutions: UK (Kindle only eBook ASIN: B009IXFJ92)

Sandle, T. and Saghee, M.R. (2013). Cleanroom Management in Pharmaceuticals and Healthcare, Euromed Communications: Passfield, UK

Sandle, T. (2013). Risk Management and Risk Assessment for Pharmaceutical Manufacturing: A contamination control perspective, Microbiology Solutions: London, UK

Sandle, T. (2013). Sterility, Sterilisation and Sterility Assurance for Pharmaceuticals: Technology, Validation and Current Regulations, Woodhead Publishing Ltd.: Cambridge, UK (ISBN 1 907568 38 7)

Pharmaceutical Microbiology

Microbiology is the study of microscopic organisms. Micro-organisms are ubiquitous and diverse, occurring in the air, sea, land, on the human skin and inside the human body. Micro-organisms include the cellular: the simple Bacteria and Archaea (prokaryotic) and more complex algae, fungi and protozoa (eukaryotic); and the non-cellular, viruses.

Micro-organisms are both beneficial and harmful to human existence. On one level bacteria are absolutely necessary for all life on this planet - for every known ecosystem - including the human ecosystem. Without bacteria, there would be no life on the Earth – from soil fertility to animal production.

On a more day-to-day level, microorganisms benefit through their use in the food industry (for example, from the production of amino acids in 'enriched' foods like bread to the creation of citric acid in soft drinks by a fungus); in medical research through the addition of human genes into bacteria ('gene expresion', for example, the manufacture of hormones like insulin) and the production of drugs and medicines ('biotechnology'). Some fungi and bacteria are capable of producing antibiotics, such as, the fungus Penicillium.

However, a minority of the thousands of microbial species are also capable of causing problems or harm to humans ('pathogenic'): from food spoilage to life threatening diseases.

Microbiology is a biological science and is made up of several sub-disciplines, these disciplines include: Immunology (the study of the immune system and how it works to protect us from harmful organisms and harmful substances produced by them); Virology, the study of viruses, and how they function inside cells; Pathogenic Microbiology, the study of disease-causing micro-organisms and the disease process and so on. Microbiology at BPL is primarily concerned with bacteria and fungi.

The microbiological discipline of relevance to healthcare and pharmaceutical manufacturing is Pharmaceutical Microbiology, an applied branch of Microbiology. This is the study of micro-organisms associated with the manufacture of pharmaceuticals, primarily in minimizing the numbers in a process environment; ensuring that the finished product is sterile and excluding those specific strains that are regarded as objectionable from starting materials and water, such as, *Escherichia coli, Salomellae, Staphylococcus aureus* and *Pseudomonas aeruginosa*. Microbiology is also concerned with toxins (microbial by-products).

These areas are examined through a number of different methods. The basis of most methods is the use of culture media, designed to cultivate and grow bacteria and fungi. The over-riding technique in each method is 'aseptic technique' (the prevention of contamination during the manipulation of samples and cultures). Both the growth of microorganisms and the maintenance of aseptic technique are complex and require trained and skilled microbiologists.

For the student, someone new to the profession, someone working in a different discipline or for the general reader, some of the language and acronyms used can be confusing (and sometimes contradictory). To aid with this, this glossary of common terms has been put together. I hope you find it of some use. Please let me have any comments, corrections or clarifications.

Tim Sandle

Email: pseudomonas@btinternet,com

Visit the Pharmaceutical Microbiology website at http://www.pharmamicroresources.com/

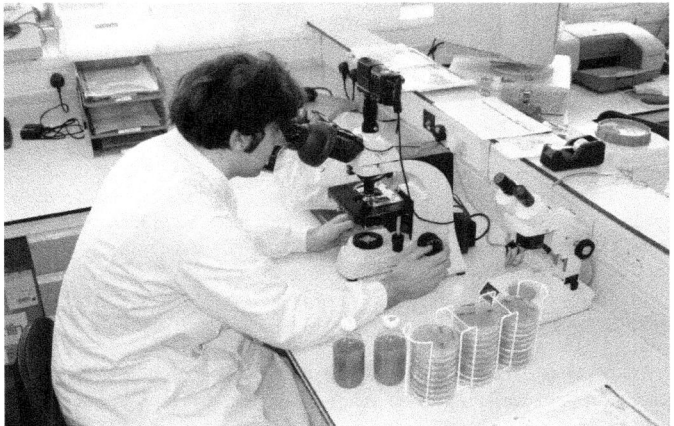

A

Accuracy - A term that indicates how closely an analytical or assay procedure approaches the true value for a particular sample. (Note that this requires knowing what the true value is.)

Acetogenic bacteria - bacteria capable of reducing CO_2 to acetic acid or converting sugars quantitatively into acetate.

Acid-fast organisms - microorganisms (e.g. mycobacteria) that resist decolorization by acid-alcohol washes that are used to remove basic stains.

Acid-fast staining - A staining procedure that differentiates between bacteria based on their ability to retain a dye when washed with an acid alcohol solution.

Acidophile- A microorganism that grows best under acid conditions (down to a pH of 1).

Activated carbon - material used to adsorb organic impurities from water. Derived from wood, lignite, pulp-mill char, blood, etc. The source material is initially charred at high temperature to convert it to carbon. The carbon is then "activated" by oxidation from exposure to high temperature steam. It comes in granular or powdered form.

Adventitious agent – Acquired, accidental contaminants

Actinomycete- a non-taxonomic term applied to a group of high G + C base composition, Gram-positive bacteria that have a superficial resemblance to fungi. Includes many but not all organisms belonging to the order Actinomycetales.

Aerobia - the plural of aerobe

Aerobic- (i) Having molecular oxygen as a part of the environment. (ii) Growing only in the presence of molecular oxygen, as in aerobic organisms. (iii) Occurring only in the presence of molecular oxygen, as in certain chemical or biochemical processes such as aerobic respiration.

Aerial hyphae - hyphae above agar surface.

Aerobiology - the study of organic particles, such as bacteria, fungal spores, very small insects, pollen grains and viruses, which are passively transported by the air.

Aerophobic - an organism harmed by oxygen; an obligate anaerobe.

Aerosol - Air suspension of solid or liquid particles having a volume median diameter of less than 50 μm. The small size of the droplets or particles allows entry to the body via the respiratory tract and readily contaminates clothing, skin and eyes.

Aerotolerant anaerobes- Microbes that grow under both aerobic and anaerobic conditions, but do not shift from one mode of metabolism to another as conditions change. They obtain energy exclusively by fermentation.

Agar- Complex sulfated polysaccharide derived from certain marine algae (normally red algae) that is a gelling agent for solid or semisolid microbiological media. Agar consists of about 70% agarose and 30% agaropectin. Agar can be melted at temperature above 100°C; gelling temperature is 40-50°C. Agar is used for performing viable counts. See also media.

Agarose - a highly purified form of agar. non-sulphated linear polymer consisting of alternating residues of D-galatose and 3,6-anhydro-L-galactose:
[-3,6-anhydro-alpha-L-galactopyranosyl-(1->3)-beta-D-galactopyranosyl-(1>4)-]n

Agarose is extracted from seaweed and is widely used as the resolving agent in electrophoresis (i.e. agarose gel electrophoresis).

Airborne concentration - The amount of particles per unit volume of air. Typically this refers to the number concentration of spray drops

Air changes – a value indicating the number of times per hour the air is changed within a certain room or containment. It is a test that indicates how often a cleanroom purges itself.

Air change rate - The number of times the total air volume of a defined space is replaced per unit of time. Typically calculated by dividing the amount of air delivered per hour by the total volume in cubic feet to give air changes per hour.

Air filter - an air cleaning device that removes particulate contamination from an airstream by straining, impingement, interception, electrostatic attraction or absorption.

Air lock - an airtight room adjoining cleanrooms or containment laboratories that acts as a buffer zone between two independent areas of unequal pressure. A pressure differential of ≥ 15 Pa is typically maintained between the inner room and the air lock; and between the air lock and the external area. Also termed Ante-room.

Air sample - the collection a sample of air; can be physical (the collection and estimation of the number of airborne particles, normally called particle counting) or viable (the estimation of the number of viable micro-organisms are either deposited onto a settle plate - passive sampling - or collected from a defined volume of air - active / volumetric air-sampling using an air-sampler.

Air shower - A relatively small, isolated "chamber" normally located at the main entrance of a cleanroom. Designed allegedly to remove gross particulate form personnel and garments by air jets.

Air velocity - a value indicating the speed of the air movement in a cleanroom or clean zone. It is an indicator of the ability of a unidirectional-flow cleanroom to purge itself.

Airborne particles - Airborne particulates are discrete particles having measurable physical boundaries in all directions and of such size and mass as to remain suspended in air long enough to be sampled and measured.

Alkalophile – a micro-organism that lives in alkaline conditions (growth range pH8.5-12.0).

Alga (plural algae) - phototrophic eukaryotic microorganisms. Algae could be unicellular or multicellular. Blue-green algae is not true algae; it belongs to a group of bacteria called cyanobacteria because it lacks a nucleus in the cell.

Ambient conditions - normal conditions, such as pressure, temperature, humidity, etc. which are considered normal for a given location.

Anaerobic- (i) Absence of molecular oxygen. (ii) Growing in the absence of molecular oxygen, such as anaerobic bacteria. (iii) Occurring in the absence of molecular oxygen, as a biochemical process.

Analyte - the specific substance to be determined in an assay or analysis.

Anemometer - an instrument for measuring air velocities.

Anoxic - without oxygen present.

Antagonistic – a micro-organism that limits or prevents the growth of another organism.

Anthrax - an infectious disease of animals caused by ingesting Bacillus anthracis spores. Can also occur in humans and is sometimes called woolsorter's disease.

Antibiotic - substances produced by the natural metabolic processes of microorganisms that can inhibit or destroy other microorganisms.

Antibiotic resistance - Plasmids generally contain genes which confer on the host bacterium the ability to survive a given antibiotic.

Antigen   - a substance that is recognized by the body as being foreign, thus it can elicit an immune response. In blood banking, antigens are usually, but not exclusively, found on the blood cell membrane.

Antimicrobial agent - a chemical that kills or inhibits the growth of microorganisms e.g. a disinfectant

Antiseptics - chemical agent used on exposed body surface to destroy or inhibit vegetative pathogens. It is the alternative name for sanitizer as in 'hand sanitizer'. The process of using such an agent is called antisepsis.

API - Analytical Profile Index - the primary method for bacterial (and some fungal) identification based on a series of individual biochemical tests, the positive and negative reactions to which form a profile which is matches for similarity against a database.

Archaea - the domain that contains procaryotes with isoprenoid glycerol diether or diglycerol tetraether lipids in their membranes and archaeal rRNA (among many differences).

Ascomycetes- A class of fungi characterized by endogenous production of spores (asocspores) in the organ of the meiosis (ascus).

Aspergillosis - a fungal disease caused by species of *Aspergillus*.

Asepsis - absence of microorganisms that cause disease; freedom of infection; exclusion of microorganisms

Aseptic technique- manipulating sterile instruments, equipment or culture media in such a way as to avoid contamination.

At rest – a term for the static state e.g. an at-rest cleanroom is a cleanroom that is complete, with all services operating but without personnel present.

Autoclave - An apparatus for sterilizing objects (destruction of microorganisms by high temperature) by the use of steam under pressure.

Autotroph – an organism that uses inorganic carbon as its sole source of carbon.

Auxotroph - an organism which cannot grow on a 'minimal' medium (e. g., mineral salts and glucose) without addition of one or more specific supplements (e. g., a specific amino acid).

B

Bacillus- (Latin for "little rod").Bacterium with an elongated, rod shape.

A genus of bacteria of the family Bacillaceae, including large aerobic or facultatively anaerobic, spore-forming, rod-shaped cells.

The term bacilli can refer to rod shaped bacteria of any genera. However the term 'rod' is more commonly used for morphological identification.

Bacteria- All prokaryotes that are not members of the domain Archaea. [**Archaea-** Evolutionarily distinct group (domain) of prokaryotes consisting of the methanogens, most extreme halophiles and hyperthermophiles, and *Thermoplasma*]. They are single celled organisms.

Bacterial Endotoxin Test – this is a pharmacopeial test for the detection of Gram-negative endotoxins using the LAL methodology.

Bactericidal - having the characteristic of destroying bacteria

Bactericide - An agent that kills bacteria. See microcidial.

Bacteriophage - a virus that exclusively infects bacteria. A protein coat surrounds the genome (DNA or RNA). One of the bacteriophages most extensively studied is the lambda phage, which is also one of the most important viral vectors used in rDNA work. Lambda promoters have been used to express eukaryotic proteins in *Escherichia coli.*

Bacteriostat - a chemical or physical agent that inhibits the growth of, but is not lethal to bacteria.

Bacteriostatic - From the Greek words Stasis and static means to stand still. A bacteriostatic agent prevents the growth of bacteria on tissue or on objects in the environment. Some disinfectants have bacteriostatic properties (as opposed to bactericidal).

Bacteriostatic Water-for-Injection - U.S.P. Water that serves the same purposes as Sterile Water for Injection, it meets the same standards, with the exception that it may be packaged in either single-dose or multiple-dose containers of not larger than 30-mL size

Bactometer - a device for the estimation of bacterial contamination within a few hours, based on measuring the early stages of breakdown of nutrients by the bacteria through changes in the electrical impedance of the medium.

Basal medium -A (n) (unsupplemented) <u>medium</u> which allows the growth of many types of microorganisms which do not require any special nutrient supplements, e.g. <u>nutrient broth</u>.

Baseline – observations or data used for comparison or as a control

Barrier technology - the technology of using separating environments, whether protecting the world from a product or the product from the world. Containment, barrier isolation and isolation all refer to the same technology, which is enclosing an environment. There are, however, some redefining terms that are gaining favor: 1. Containment - protect the world from the product (as in the case of highly potent compounds or a toxic). 2. Isolation - protect the product from the world (as in the case of a sterile product). 3. ISO 14644-7 "Minienvironments and Isolators" will define further levels of devices

Batch - a specific quantity of material produced in a process or series of processes so that is expected to be homogeneous within specified limits. In the case of continuous production a batch may correspond to a defined fraction of the production, characterized by its intended homogeneity. The batch size may be defined either by fixed quantity or the amount produced in a fixed time interval.

Batch culture –a microbial laboratory culture grown in a fixed volume of growth medium; designed to produce the maximum density.

Benthos – micro-organisms living at the bottom of an aquatic system, associated with sediments.

Bioassay - the determination of the biological activity of a substance (e.g. a drug) by observing its effect on an organism (or organ) compared to a standard preparation.

Bioburden – a 'catch all phrase' that refers to the total microbial load of a sample or system. It normally infers unwanted contamination but can refer to the 'natural bioburden' or a substance.

Bioburden assay – an alternative name for a dilution series or test designed to enumerate the microbial content of a sample.

Biocide - an agent that can kill all pathogenic and non-pathogenic living organisms, including spores. More general than bactericide, biocide includes insecticides and any compound toxic to any living thing.

Biochemistry - The study of those molecules used and manufactured by living things.

Biochemical oxygen demand (BOD) - the amount of oxygen used by organisms in water under certain standard conditions; it provides an index of the amount of microbially oxidizable organic matter present.

Biodegradable - material that can be broken down by biological action.

Biological Indicator - a device used to validate items being sterilized through a sterilization procedure and to monitor adequacy of sterilization. The device consists of a known number of microorganisms (usually bacterial spores), of known resistance to the mode of sterilization, in or on a carrier and enclosed in a protective package. Subsequent growth or failure of the microorganisms to grow under suitable conditions indicates the adequacy of sterilization.

Biofilm- Microbial cells encased in an adhesive, usually a polysaccharide material secreted by the community, and attached to a surface. They have complex structural and functional characteristics. Biofilms have physical/chemical gradients that influence microbial metabolic processes. They can form on inanimate devices (e.g. pipes) and also cause fouling (e.g. water systems). Often occurs at a physical e.g. water – solid interface.

Biogenerator - a contained system, such as a fermentor, into which biological agents are introduced along with other materials so as to effect their multiplication or their production of other substances by reaction with the other materials. Biogenerators are generally fitted with devices for regulation, control, connection, material addition, and material withdrawal.

Biohazard - an infectious agent(s), or part thereof, presenting a real or potential risk to human, other animals, or plants, directly through infection or indirectly through disruption of the environment.

Bioluminescence- the production of light by a chemical reaction within an organism. The process occurs in many bacteria and protists, as well as certain animals and fungi.

Biomass – the total mass of all living organisms, or set of organisms, usually expressed as a dry weight.

Biome – a community of micro-organisms living within an ecosystem.

Binomial system - the nomenclature system in which an organism is given two names; the first is the capitalized generic name, and the second is the uncapitalized specific epithet.

Bioremediation - the use of biologically mediated processes to remove or degrade pollutants from specific environments. Bioremediation can be carried out by modification of the environment to accelerate biological processes, either with or without the addition of specific microorganisms.

Biosafety cabinet – another term for microbiological safety cabinet.

Biota – living organisms.

Bioterrorism - the intentional or threatened use of viruses, bacteria, fungi, or toxins from living organisms to produce death or disease in humans, animals, and plants.

Blastospore - a spore formed by budding from a hypha.

Blotting - A technique for detecting one RNA within a mixture of RNAs (a Northern blot) or one type of DNA within a mixture of DNAs (a Southern blot). A blot can prove whether that one species of RNA or DNA is present, how much is there, and its approximate size. Basically, blotting involves gel electrophoresis, transfer to a blotting membrane (typically nitrocellulose or activated nylon), and incubating with a radioactive probe. Exposing the membrane to X-ray film produces darkening at a spot correlating with the position of the DNA or RNA of interest. The darker the spot, the more nucleic acid was present there.

BOD (Biological Oxygen Demand) - The oxygen used in meeting the metabolic needs of aerobic organisms in water containing organic compounds.

BSE - Bovine Spongiform Encephalopathy. A neurological disease of cattle which is generally thought to have caused the incidence of vCJD in humans.

Broad Spectrum - over a wide range. A broad-spectrum disinfectant is effective against a wide range of microorganisms including bacterial spores, mycobacteria, non-lipid and lipid viruses, fungi, and vegetative bacteria.

Broad-spectrum drugs - chemotherapeutic agents that are effective against many different kinds of pathogens.

Brownian motion - The random movement of molecules which is increased as energy levels are increased (heat added) and decreased as energy levels are decreased (heat removed).

Budding - a vegetative outgrowth of yeast and some bacteria as a means of asexual reproduction; the daughter cell is smaller than the parent.

Buffer - A substance capable of neutralizing both acids and bases in solution, thereby maintaining the original acidity or causticity of the solution.

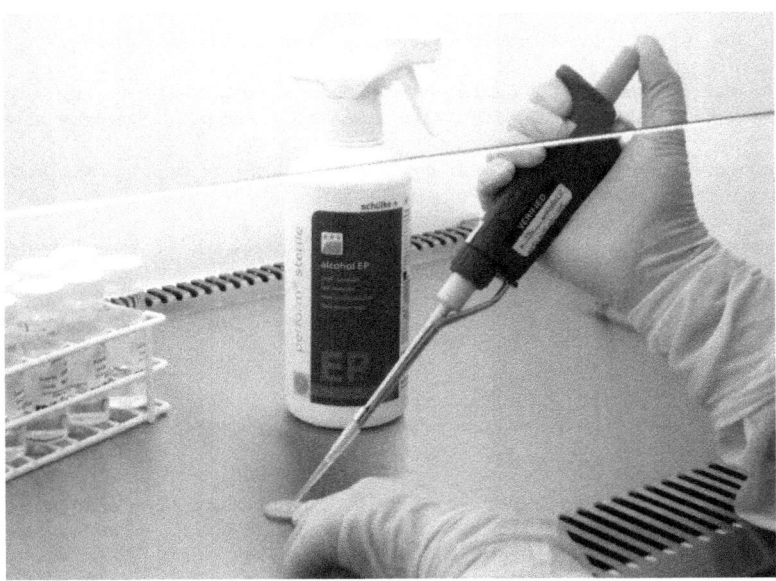

C

Calibration - Comparison of a measurement standard or instrument of unknown accuracy with another standard or instrument of known accuracy to detect, correlate, report, or eliminate by adjustment, any variation in the accuracy of the unknown standard or instrument.

Capsule- compact layer of polysaccharide exterior to the cell wall in some bacteria

Cell - The fundamental unit of life. The living tissue of almost every organism is composed of these fundamental living units. Unicellular organisms, such as yeast or a bacterium, perform all life functions within the one cell. In a higher organism, a multicellular organism, entire populations of cells may be designated a particular task. The cells of muscle tissue, for example, are specialized for movement.

Cell wall- layer or structure that lies outside the cytoplasmic membrane; it supports and protects the membrane and gives the cell shape

Centrifugal air sampler – one of two commonly use viable air-samplers. It collects micro-organisms suspended in air through centrifugal forces which 'forces' micro-organisms out of the air stream and onto microbiological media. See **impaction air-sampler.**

CFR – Code of Federal Regulations. US regulations. Parts 210 and 211 apply to biologics.

Chemolithotrophic autotrophs - microorganisms that oxidize reduced inorganic compounds to derive both energy and electrons; CO_2 is their carbon source. Also called chemolithoautotrophs.

Chemoorganotrophic heterotrophs- Microorganisms that use organic compounds as sources of energy, hydrogen, electrons, and carbon for biosynthesis.

Chemostat – a growth chamber that keeps a bacterial culture at a specific volume and rate of growth by limiting nutrient medium and removing spent culture.

chemotaxis – response of motile micro-organisms, by movement towards or away from, a chemical source.

Chitin - a tough, resistant, nitrogen-containing polysaccharide forming the walls of certain fungi, the exoskeleton of arthropods, and the epidermal cuticle of other surface structures of certain protists and animals.

CIP – Clean-in-place. A way to clean large vessels (tanks, piping and associated equipment) without moving them or taking them apart, using high pressure rinsing equipment. Sometimes followed by SIP (steam-in-place) sanitization steps.

Chlamydospore - an asexually produced, thick-walled resting spore formed by some fungi.

Chloramphenicol - a broad-spectrum antibiotic that is produced by *Streptomyces venezuelae* or synthetically; it binds to the large ribosomal subunit and inhibits the peptidyl transferase reaction.

Clarify – a process whereby a liquid is cleared of particles or micro-organisms by filtration or centrifugation.

Classification- (i) Arrangement of organisms into groups based on mutual similarity or evolutionary relatedness; (ii) pertaining to the certification of ca cleanroom.

clean-air device - a laminar flow / unidirectional flow enclosure, clean bench, clean work station, wall or suspended ceiling module, or other device (except a cleanroom) which incorporates a HEPA filter(s) and a fan(s) to supply laminar flow clean air to a controlled work space.

Cleanroom – a room in which the concentration of airborne particles is controlled to a defined standard. This is achieved by controlling the introduction, formation and retention of particles. Controlled limits may be set for air flow patterns, air cleanliness, viable and non-viable airborne particles, temperature and relative humidity, air pressure, and operating procedures.

Cleanroom facility - As Built: A cleanroom facility that is complete and ready for operation, with all services connected and functional, but without production equipment or personnel.

Cleanroom facility - At Rest: A cleanroom facility that is complete with production equipment installed, but without personnel.

Cleanroom facility - In Operation: A cleanroom facility in normal operation, with all services functioning and with production equipment and personnel present.

Cleanroom garments - special clothing designed to protect cleanroom environments from contaminants released by workers. Special apparel includes non-shedding gowns or coveralls, head covers, face masks, gloves, footwear or shoe covers.

Clean zone – a defined space within a cleanroom where the concentration of airborne particles is controlled. Often this is to a higher standard than the cleanroom.

Clone (verb) - To "clone" something is to produce copies of it.

Clone (noun) - The term "clone" can refer either to a bacterium carrying a cloned DNA, or to the cloned DNA itself.

Coagulase - an enzyme that induces blood clotting; it is characteristically produced by pathogenic *staphylococci.*

Coccoid - sphere-shaped

Coccus- Spherical bacterial cells

Coliform- Gram-negative, non-spore-forming facultative rod that ferments lactose with gas formation with 48 hours at 35°C. Often an indicator organism for fecal contamination of water supplies. *Escherichia coli* and *Enterobacter are* important members

Colonization - multiplication of a microorganism after it has attached to host tissues or other surfaces.

Colony - a cluster or assemblage of microorganisms growing on a solid surface such as the surface of an agar culture medium; the assemblage often is directly visible, but also may be seen only microscopically.

Colony forming unit (CFU) - Colony forming units. Viable micro-organisms (bacteria, yeasts & mould) capable of growth under the prescribed conditions (medium, atmosphere, time and temperature) develop into visible colonies (colony forming units) which are counted. The term colony forming unit (CFU) is used because a colony may result from a single micro-organism or from a clump / cluster of micro-organisms. It is normally expressed as CFU per g or mL.

Column – a vertical, cylindrical container or vessel used in separation processes including extraction, distillation or chromatography.

Community- All organisms that occupy a common habitat and interact with one another. A biofilm, for example, is a type of community.

Compendial - official; purported to comply with a pharmacopeia.

Conidiospore - an asexual, thin-walled spore borne on hyphae and not contained within a sporangium; it may be produced singly or in chains.

Contact plate - see RODAC

Containment - The utilization of filters and controlled ventilation systems in a designated space for capturing and containing potential contaminants.

Contaminant – a foreign agent not introduced as part of processing, such as airborne particulates or adventitious micro-organisms.

contaminated - possessing infectious organisms or substances

Control Area - A building or portion of a building within which the exempted amounts of hazardous materials may be stored, dispensed, handled, or used.

Culture- Population of microorganisms cultivated in an artificial growth medium. A pure culture is grown from a single cell; a mixed culture consists of two or more microbial species or strains growing together

Culture collection - this refers to the storage and preservation of pure cultures of microorganisms on suitable survival media. Such microorganisms are normally stored by cryopreservation (<-70°C). There are several international culture collections from which typed strains of pure microorganisms are purchased (for example, the American Type Culture Collection - ATCC - or the UK National Typed Culture Collection - NTCC.

Culture medium - an aqueous solution of various nutrients suitable for the growth of microorganisms. See media.

Cyst - a general term used for a specialized microbial cell enclosed in a wall. Cysts are formed by protozoa and a few bacteria. They may be dormant, resistant structures formed in response to adverse conditions or reproductive cysts that are a normal stage in the life cycle.

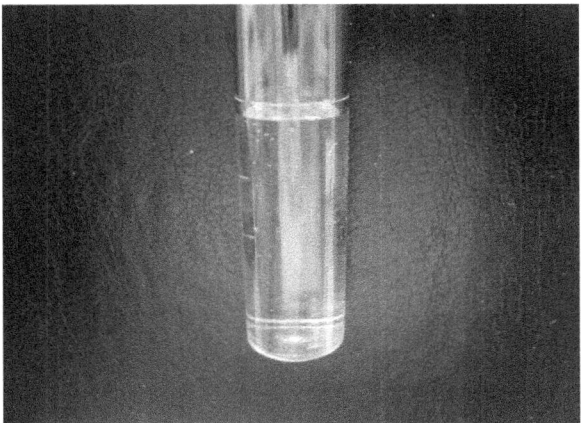

D

Dark-field microscopy - microscopy in which the specimen is brightly illuminated while the background is dark. This can be useful for discerning certain colonies.

Death phase - the decrease in viable microorganisms that occurs after the completion of growth in a batch culture.

Decimal reduction time (D or D value) The time required to kill 90% of the microorganisms or spores in a sample at a specified temperature.

Decontamination - to make safe by removing or reducing contamination by infectious organisms or other harmful substances. It does not imply that all micro-organisms have been eliminated to zero count.

Defined medium - culture medium made with components of known composition.

Degradation – loss or reduction of quality, integrity or character; or a reaction that breaks something down into smaller parts.

Deionized water - Also known as DI water: Water from which the salts and ions usually contained in water have been removed.

Dendrogram - a treelike diagram that is used to graphically summarize mutual similarities and relationships between microorganisms.

Deoxyribonucleic acid (DNA) - the nucleic acid that constitutes the genetic material of all cellular organisms. It is a polynucleotide composed of deoxyribonucleotides connected by phosphodiester bonds.

Depyrogenation – the process for the removal of pyrogenic substances, normally using chemical or thermometric processes. It is measured using Gram-negative endotoxin as a heat resistant challenge but the process refers to the removal of all pyrogens,

Detergent - an organic molecule, other than a soap, that serves as a wetting agent and emulsifier; it is normally used as cleanser, but some may be used as antimicrobial agents

Differential medium- Cultural medium with an indicator, such as a dye, which allows various chemical reactions to be distinguished during growth. The medium allows growth of several microbes but creates different appearances. An example is MacConkey agar.

Diluent – a chemically inert substance added to a solution to increase the volume and to reduce the concentration; a diluting agent.

Dilution plate count method- Method for estimating the viable numbers of microorganisms in a sample. The sample is diluted serially and then transferred to agar plates to permit growth and quantification of colony-forming units

Direct count- Method of estimating the total number of microorganisms in a given sample by direct microscopic examination

Direct inoculation - the secondary sterility test method and a crude technique to examine for microbial growth by transferring a portion of a liquid or solid into a broth media, incubating for a defined time, and then examining for observable growth.

Discrete particle counter - alternative name for a **particle counter**.

Disease - a deviation or interruption of the normal structure or function of any part of the body that is manifested by a characteristic set of symptoms and signs.

Disinfectant - an agent that kills all growing or vegetative forms of microorganisms, thus completely eliminating them from inanimate objects.

Disinfection - refers to the use of a physical process or a chemical agent (such as disinfectants) to destroy vegetative pathogens but not necessarily bacterial endospores.

Disinfection by-products (DBPs) - Chemical released or created after the use of a disinfectant. For example, chlorinated organic compounds formed during chlorine use for water disinfection. Many are carcinogens.

Domain – a major taxonomic assemblage.

Dosage form - a pharmaceutical product as produced for use (tablet, capsule, etc.)

Doubling time - the time needed for a population to double.

DQ – design qualification. A documented review of the design, at appropriate stages of a project, for conformance to operational and regulatory expectations.

Dynamic state – refers to environmental or particle monitoring where a room is occupied (in less politically correct times this was called the 'manned state').

Dynamic contamination - occurs when organisms are actively metabolizing. The numbers are increasing (10E4 or >) and the organism is usually a Gram-negative bacterial species. Only the dynamically contaminated product can be spoiled.

E

Ecosystem – a self-regulating biological community in a defined habitat

Efficacy – the ability of a substance (e.g. chemical disinfectant) to produce a desired effect (e.g. reduction in bacterial numbers) under different conditions.

Elution – washing out; removal of material; the separated material is the eluate.

Emulsification – a process that creates a stable mixture of two solutions that would not normally mix together (such as oil and water) by forcing one to disperse with the other as droplets.

Endemic species – micro-organisms restricted to a certain location

Endogenous – growing or developing from within a micro-organism

Endospore- Differentiated cell formed within the cells of certain Gram-positive bacteria and extremely resistant to heat and other harmful agents. Sometimes simplified to 'spore'.

Endotoxin - a toxin, from the outer cell membrane of Gram-negative bacteria, consisting of heat stable lipopolysaccharide (LPS) molecules. It is normally released by cell lysis. Can cause septic shock and tissue damage in people.

Enrichment culture- Technique in which environmental (including nutritional) conditions are controlled to favor the development of a specific organism or group of organisms.

Enriched media - A medium which ensures growth of a particular microbe. An example is blood agar.

Enteric bacteria - members of the family Enterobacteriaceae (gram-negative, peritrichous or non-motile, facultatively anaerobic, straight rods with simple nutritional requirements); also used for bacteria that live in the intestinal tract.

Enterobacteriaceae / enteric bacteria- General term for a group of bacteria that inhabit the intestinal tract of humans and other animals. Among this group are pathogenic bacteria such as *Salmonella* and *Shigella* and consists of members of the family Enterobacteriaceae (gram-negative, peritrichous or non-motile, facultatively anaerobic, straight rods with simple nutritional requirements).

Environmental monitoring – a documented series of sampling and testing in controlled environments in order to demonstrate conformance to a series of pre-set limits or for trends. It is distinct from environmental control.

Enzyme - Any of numerous proteins or conjugated proteins produced by living organisms and functioning as complex biochemical catalysts. They not only promote reactions but also function as regulators making sure the organism does not produce too much or too little of any chemical substance. Although all enzymes are proteins, many contain additional non-protein components essential for catalytic activity. Such enzymes are termed haloenzymes. The protein part of this enzyme is termed an apoenzyme and the non-amino acid part is termed a coenzyme.

Epidemic - a disease that suddenly increases in occurrence above the normal level in a given population.

Eucaryotic cells - cells that have a membrane-delimited nucleus and differ in many other ways from procaryotic cells; protists, algae, fungi, plants, and animals are all eucaryotic.

Excipient – a raw material that is present in a drug product and thus has direct patient contact. Includes bulking agents, stabilizing agents, preservatives, salts, solvents or waters.

Exogenous – developing from the outside; can include external factors that affect the growth or survival of a micro-organisms, such as light, humidity, nutrients etc.

Exotoxins - poisonous substances produced by the microbial cell and liberated into the surrounding environment, without destruction of the cell.

Exponential growth- Period of sustained growth of a microorganism in which the cell number constantly doubles within a fixed time period.

Exponential phase- Period during the growth cycle of a population in which growth increases at an exponential rate. As referred to as logarithmic phase

Extractables – substances that are withdrawn from a container or closure through a process.

Extreme environment – an environment in which physicochemical factors (e.g. pH, temperature, salinity etc.) are outside of the normal range for most micro-organisms

Extremophiles - microorganisms that grow under harsh or extreme environmental conditions such as very high temperatures or low pHs. In relation, an extreme environment is one in which physical factors such as temperature, pH, salinity, and pressure are outside of the normal range for growth of most microorganisms; these conditions allow unique organisms to survive and function.

F

f-value - the time required to reduce the total number of micro-organisms in a certain medium to a required level.

Facultative anaerobes - microorganisms that do not require oxygen for growth, but do grow better in its presence.

Facultative organism- Organism that can carry out both options of a mutually exclusive process (e.g., aerobic and anaerobic metabolism).

FDA-483 – a form prepared by at the conclusion of an inspection by the Federal Drug Administration (FDA), which cite observations that may constitute violations of federal law (in the opinion of the inspector).

Fermentation - The process of growing microorganisms within an enclosed tank (fermenter) under controlled conditions of aeration, agitation, temperature, and pH. The different types organisms used as a basis for fermentation are:

1. Bacteria (E. coli)
2. Yeasts
3. Molds
4. Chinese Hamster Ovary (CHO) cells
5. Kidney cells
6. Vaccines to viruses

Fiber - a particle having a length 100 micrometers or greater, and an aspect ratio of at least 1:10

Filamentous- In the form of very long rods, many times longer than wide (for bacteria), in the form of long branching strands (for fungi).

Filter – a porous material through which a liquid or gas is passed so that particulates and impurities and micro-organisms are held in suspension. Some 'impurities' will pass through depending upon the filter size, viscosity, concentration and so on. The part of the mixture that passes through is the filtrate.

Filtration - Removal of suspended matter from a fluid by passing it through a porous matrix that prevents particles from getting through, usually by entrapment on or in the filter matrix.

Filtration rate – the volume of solution cleared of particles per unit time.

Fimbria (plural, fimbriae) - Short filamentous structure on a bacterial cell; although flagella-like in structure, generally present in many copies and not involved in motility. Plays a role in adherence to surfaces and in the formation of pellicles.

Fixation - the process in which the internal and external structures of cells and organisms are preserved and fixed in position. As with the Gram-stain.

Flagellum (plural, flagella) - Whip like tubular structure attached to a microbial cell responsible for motility

Fluorescent- having the ability to emit light of a certain wavelength when activated by light of another wavelength

FMEA – Failure Modes Evaluation and Analysis (or Failure Mode and Effects Analysis). A risk assessment and risk mitigation tool. Similar to HACCP.

Food-borne infection - gastrointestinal illness caused by ingestion of microorganisms, followed by their growth within the host. Symptoms arise from tissue invasion and/or toxin production.

Formulation – the method and process for selecting the components of a mixture, including the active ingredients, to make the form that the drug is finally administered in (such as a tablet or injection)

Fractional sterilization - a method of sterilization that involves alternating exposure and cooling time for a consecutive period.

Fruiting body- Macroscopic reproductive structure produced by some fungi, such as mushrooms, and some bacteria, including myxobacteria. Fruiting bodies are distinctive in size, shape, and coloration for each species

Fumigant- liquid or solid chemical that forms vapors that kill organisms.

Fungicide- chemical or physical agent that kills or inhibits development of fungus spores or mycelium. The term 'fungicide' includes all preparations intended for preventing, destroying, repelling or mitigating any fungi.

Fungistasis- Suppression of germination of fungal spores or other resting structures in natural soils as a result of competition for available nutrients, presence of inhibitory compounds, or both.

Fungistasis- The prevention of fungal growth; the effect is reversible, if the inhibitor is removed or diluted, growth is resumed.

Fungistatic - Able to inhibit germination of fungus spores or development of mycelium without causing death of fungus.

Fungus (plural, fungi) - Non-phototrophic, eukaryotic microorganisms that contain rigid cell walls. An example of microscopic fungi are yeasts and molds and macroscopic fungi are mushrooms, puff balls and bracket fungi. All fungi are heterotrophic; living off of organic substrates (dead animal/plant material), or on the tissue of other organisms (parasite). Their reproduction is primarily through spore formation.

G

Gel electrophoresis - A method to analyze the size of DNA (or RNA) fragments. In the presence of an electric field, larger fragments of DNA move through a gel slower than smaller ones. If a sample contains fragments at four different discrete sizes, those four size classes will, when subjected to electrophoresis, all migrate in groups, producing four migrating "bands". Usually, these are visualized by soaking the gel in a dye (ethidium bromide) which makes the DNA fluoresce under UV light.

General purpose media - Medium to grow as wide a variety of microbes as possible. An example is TSA

Generation time – the time required for a microbial population to double in number

Genotype- Precise genetic constitution of an organism.

Genus (plural, genera) - The first name of the scientific name (binomial); the taxon between family and species

Geobacillus stereathermophillus - a highly heat-resistant, endospore-forming microorganism used in form on commercially prepared spore strips for checking sterilization effectiveness in steam sterilizers.

Germicide - An agent that destroys microorganisms, especially pathogenic microorganisms ("germs"). Sterilants, disinfectants, and antiseptics are germicides.

Germination - the stage following bacterial endospore activation in which the endospore breaks its dormant state. Germination is followed by outgrowth.

Glovebox - an enclosure, fitted with sealed gloves that allows external manual manipulations in controlled or hazardous environments. Applications include nuclear, biomedical, semiconductor, chemical industries; and research laboratories - where isolation of the controlled zone is required.

Glucans - polysaccharides composed of glucose units held together by glycosidic linkages. Some types of glucans have $\beta(1\_3)$ and $\beta(1\_6)$ linkages and bind bacterial cells together on teeth forming a plaque ecosystem. Can cause a positive response in the LAL assay and can therefore sometimes be mistaken for endotoxin.

Glycocalyx - the means by which a microbial community adheres as a biofilm. It is a coating of macromolecules which protects some kinds of bacteria. It helps bacteria adhere to its environment. It can differ in thickness, organization and chemical composition. Two types are slime layer and capsule

Good Manufacturing Practice (GMP) and current Good Manufacturing Practice (cGMP) – standards which relate to the manufacturing of biopharmaceuticals and blood products. Include SOPs to be followed; processes to be validated; equipment to be qualified; staff to be trained; and a clean environment to be maintained.

Gram-negative cell - a bacterial cell whose cell wall contains relatively little peptidogylcan but has an outer membrane composed of lipopolysaccharide, lipoprotein, and other complex macromolecules. Gram-negative and positive micro-organisms are distinguished using the Gram stain. Gram-negative micro-organisms do not retain the primary stain (crystal violet) and require counter-staining.

Gram-positive cell - a bacterial cell whose cell wall consists mainly of peptidogylcan and lacks the outer cell membrane found on Gram-negative cells. Gram-positive micro-organisms retain the primary stain (crystal violet).

Gram's stain- Differential stain that divides bacteria into two groups, Gram-positive and Gram-negative, based on the ability to retain crystal violet when decolorized with an organic solvent such as ethanol. The cell wall of Gram-positive bacteria consists chiefly of peptidoglycan and lacks the outer membrane of Gram-negative cells. The Gram-stain is the basis of bacterial identification. Gram-negative and positive micro-organisms are distinguished using the Gram stain.

Growth - an increase in cellular constituents.

Growth factors - organic compounds that must be supplied in the diet for growth because they are essential cell components or precursors of such components and cannot be synthesized by the organism.

Growth rate- The rate at which growth occurs, usually expressed as the generation time

GXP – All-inclusive term for Good Clinical Practice, Good Laboratory Practice and Good Manufacturing Practice.

H

Habitat- the place where an organism lives. The habitat typically contains all the conditions to either simulate growth or to preserve the organism.

Halobacteria or extreme halophiles - a group of archaea that have an absolute dependence on high NaCl concentrations for growth and will not survive at a concentration below about 1.5 M NaCl.

Halophile- Organism requiring or tolerating a saline environment.

Hazard Analysis and Critical Control Points (HACCP) - a method of determining the hazards in a process and to control them.

HEPA - High Efficiency Particulate Air - describes the system for filtering (diluting) air into cleanrooms. Standard HEPA filters remove 99.97% of 0.3μm particles.

Heterotroph- Organism capable of deriving carbon and energy for growth and cell synthesis from organic compounds; generally also obtain energy and reducing power equivalents from organic compounds.

Holoplanktonic – micro-organisms found in a water system for most of the year

HVAC - Heating, Ventilation and Air Conditioning - a system for providing air into clean rooms and controlling their classification by removing airborne particles through the use of HEPA filters.

Hybridization - The reaction by which the pairing of complementary strands of nucleic acid occurs. DNA is usually double-stranded, and when the strands are separated they will re-hybridize under the appropriate conditions. Hybrids can form between DNA-DNA, DNA-RNA or RNA-RNA. They can form between a short strand and a long strand containing a region complementary to the short one. Imperfect hybrids can also form, but the more imperfect they are, the less stable they will be (and the less likely to form). To "anneal" two strands is the same as to "hybridize" them.

Hydrology – the study of the inflow and outflow of water

Hydrophilic – having an affinity for water; dissolving in water

Hydrophobic – insoluble in water; resisting or repelling water

Hypha (pl., hyphae) - the unit of structure of most fungi and some bacteria; a tubular filament.

I

Immune system - the defensive system in a host consisting of the nonspecific and specific immune responses. It is composed of widely distributed cells, tissues, and organs that recognize foreign substances and microorganisms and acts to neutralize or destroy them.

Impaction air-sampler – one of two commonly used air-samplers. Air is drawn into an air-sampler and impacts onto the surface of a culture medium . The force of the impaction will deposit any micro-organisms onto the agar surface. See **centrifugal air-sampler**.

Incubation - the process of maintaining appropriate conditions (temperature, humidity, time) to allow microbial cells to replicate.

Indicator organism - an organism whose presence indicates the condition of a substance or environment, for example, the potential presence of pathogens. This is sometimes called a 'specified' or 'objectionable' microorganism and they are examined for in raw materials and in some water samples. For example, in water, coliforms are used as indicators of faecal pollution.

Infection - invasion of the body by pathogenic microorganisms, and the reaction of tissues

In-house – work conducted in the laboratory; not out sourced

Inoculate- To transfer a liquid, product or micro-organism from one container or agar medium to another. See 'direct inoculation'.

Inoculum- Material used to introduce a micro-organism into a suitable situation for growth.

Ionizing radiation - radiation of very short wavelength or high energy that causes atoms to lose electrons or ionize

Intermediates / in-process – substances formed or parts of the middle of a process that make up a series of processing steps between raw materials and the final products.

In vitro- Literally "in glass"; it describes whatever happens in a test tube or other receptacle, as opposed to *in vivo*. When a study or an experiment is done outside the living organism, in test tube, it is done *in vitro*.

In vivo- In the body, in a living organism, as opposed to *in vitro*; when a study or an experiment is done in the living organism, it is done *in vivo*

ISO Cleanroom Classification - Cleanroom air cleanliness classifications that provide standards for the identification, control, and monitoring of various airborne particles and contaminants. The series falls under the title ISO 14644 for physical parameters and ISO 14698 for biocontamination.

Isokinetic – the taking of a sample where the air entering the sampler is at the same velocity and direction as the air in the cleanroom (**isoaxial sampling**). The opposite term is anisokinetic.

Isolation- Any procedure in which an organism present in a particular sample or environment, is obtained in pure culture.

Isolator – Self-contained clean spaces designed to protect product from contamination by room that exchange air with the surrounding environment only if the air has passed through a filter of at least HEPA quality.

IQ – **installation qualification**. Documented verification that aspects of a facility, utility or equipment meet approved specifications and have been correctly installed.

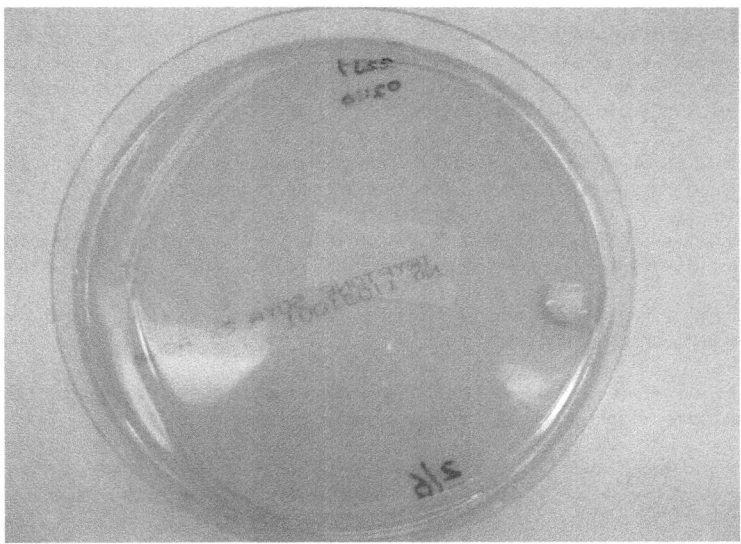

J,K,L

Joule - a unit of work or energy equal to 10,000,000 ergs; 1000 joules equals 1 kilojoule (kJ).

Kingdom - the highest category in the taxonomic hierarchy of classification.

Lag phase - the early period of growth undergone by a microbial cell. Growth is slow following inoculation into a culture medium. See **logarithmic phase**.

Laminar Airflow - where air travels in either a vertical or horizontal direction at the *same* velocity. 'Laminar air flow' is a misnomer; more correctly termed 'unidirectional flow'.

LAF / LAC - Laminar airflow cabinets are designed to direct particles away from the work surface (this is often supported by airflow or smoke studies). The term 'LAF' is becoming increasingly replaced by UDAF or uni-directional air-flow. Thee terms have different meanings (see **UDAF** and **turbulent flow** below)

Latent phase – the period following the introduction of micro-organisms into a fresh culture medium when there is no increase in biomass or population count.

Leachable – chemical entity that has the potential to be extracted from a container or closure when exposed to certain conditions or solutions.

Lichen - an organism composed of a fungus and either green algae or cyanobacteria in a symbiotic association.

Limit of detection - the lowest concentration of an analyte that can be detected reliably (present or absent) in a particular sample. Exact definitions vary.

Limit of quantitation - the lowest concentration of an analyte that can be determined quantitatively (at acceptable precision) in a particular sample. Exact definitions vary.

Limit tests - typically qualitative tests which show whether the concentration of a particular substance is above or below the pharmacoepial limit.

Microbial limit tests are usually for total aerobic count, or for presence of *Staphylococcus aureus, Pseudomonas aeruginosa, Salmonella*, or *Escherichia coli* in substances which are not required to be sterile.

***Limulus* amebocyte lysate** - LAL, the biochemical method used for the Bacterial Endotoxin Test. There are three methods: gel-clot, turbidimetric and chromogenic, based on an extract of the blood of the crab *Limulus polyphemus.*

Lipopolysaccharide (LPS)-Complex lipid structure containing unusual sugars and fatty acids found in many Gram-negative bacteria. It is the basis of endotoxin.

LPS-binding protein - a special plasma protein that binds bacterial lipopolysaccharides and then attaches to receptors on monocytes, macrophages, and other cells. This triggers the release of IL-1 and other cytokines that stimulate the development of fever and additional endotoxin effects.

Logarithmic (log) phase - the stage of active microbial cell division.

Lyophilization – freeze-drying; a procedure by which a liquid solution is frozen to a glassy slate (primary drying), then slightly heated to remove unfrozen water by sublimation.

Lysis- A breakdown or dissolution of cells resulting in loss of cell contents.

M

Mats (cleanrooms) - strippable fabric mesh or plastic tacky mats used at entrances to clean environments early in clean construction operations during times plastic tacky mats become loaded up with contamination so rapidly that they may not be practically maintained. An alternative is polymeric flooring.

Medium (plural, media)- Any liquid or solid material prepared for the growth, maintenance, or storage of microorganisms

Media simulation trial (or **broth filling trial**) – the exposure of microbiological growth medium to product contact surfaces, container closure systems, critical environments and process manipulations to closely simulate the exposure that product itself will undergo.

Membrane filter technique The use of a thin porous filter made from cellulose acetate, cellulose nitrate or some other polymer (normally 0.45μm) to collect microorganisms from a liquid or air sample. It is the primary sterility testing method.

Menstrumm – a solvent (as in a raw material).

Mesophile- Organism whose optimum temperature for growth falls in an intermediate range of approximately 15 to 40°C.

Metabolism - the total of all chemical reactions in the cell; almost all are enzyme catalyzed.

Microaerophile- Organism that requires a low concentration of oxygen for growth. Sometimes indicates an organism that will carry out its metabolic activities under aerobic conditions but will grow much better under anaerobic conditions.

Microbial ecology - the study of microorganisms in their natural environments, with a major emphasis on physical conditions, processes, and interactions that occur on the scale of individual microbial cells.

Microbial population- Total number of living microorganisms in a given volume or mass.

Microbiocide
- The root -cide meaning to kill can be combined with other terms to define an antimicrobial agent aimed at destroying a certain group of microorganisms.
- Bactericide is a chemical that destroys bacteria except for those in the Endospore stage.
- Fungicide is a chemical that can kill fungal spores, hyphae, and yeast.
- Virucide is any chemical known to inactivate viruses, especially on living tissue.
- Sporicide is an agent capable of destroying bacterial endospores .

Microbiology- Study of microorganisms

Microbiological safety cabinet - Bio Safety Hood Classification

- The Class I Biosafety cabinet is a ventilated cabinet with in inward airflow and outlet HEPA filters. It was previously referred to as the CDC Hood and served a valuable function in its time by protecting personnel and environment. Because it offers no product protection, it has been essentially obsolete for the past several decades.

- Class II (Types A and B) are Laminar Flow Biological Safety Cabinets that protect personnel, product and environment. They provide inward airflow to protect personnel, downflow HEPA filtered air to the room area to protect the product and exhaust HEPA filtered air to protect the environment from particulate and aerosol hazards.

- Class III cabinet is defined as a ventilated glovebox. This is a gas-tight chamber operated through sealed gloves which provide a complete barrier between the worker and hazardous material. The glovebox is maintained under negative pressure with HEPA filtered supply air and double HEPA filtered exhaust air.

Microenvironment – a small-scale habitat within a general ecosystem

Micrometre - see Micron.

Micron – a micrometre (1/100th of a millimetre or 1/1,000,000th of a meter)

Microorganism - a microscopic organism consisting of a single cell or cell cluster including the viruses.

Microcosm- A community or other unit that is representative of a larger unity.

Microenvironment- Immediate physical and chemical surroundings of a microorganism

Microflora- Bacteria (including actinomycetes), fungi, algae, and viruses found in a given location. For the human body (and related to this, the majority of the micro-organisms found in the Aseptic Filling Suite), the term **Normal Flora** can be applied. It can also called resident flora and consists of a variety of microorganisms adapted to the body at the sites of the human body that harbor normal flora are skin, digestive tract, upper respiratory tract, external genitourinary tract and external surfaces of the eyes and ears. The majority of such micro-organisms isolated are *Staphylococci* and *Micrococci*.

Microhabitat- Clusters of microaggregates with associated water within which microbes function. May be composed of several microsites (e.g., aerobic and anaerobic).

Microorganism (microbe)- Living organism too small to be seen with the naked eye (< 0.1 mm); includes bacteria, fungi, protozoans, microscopic algae, and viruses.

Minimal inhibitory concentration (MIC)- the lowest concentration of a drug that will prevent the growth of a particular microorganism.

Minimal lethal concentration (MLC) - the lowest concentration of a drug that will kill a particular microorganism.

Membrane filter - porous membrane composed of pure and biologically inert cellulous esters, polyethylene, or other materials.

Most probable number - the MPN statistical technique consists of a variety of dilution broths to test a material for the number of viable microorganisms where inhibition of the material on microbial growth maybe a problem. The broths are examined for the 'endpoint' (where growth no longer occurs) to estimate the microbial number. Expressed as a density or population of organisms per 100 mL of sample water

Motility- Movement of a cell under its own power

Mold- any of a large group of fungi that cause mould or moldiness and that exist as multicellular filamentous colonies; also the deposit or growth caused by such fungi. Moulds typically do not produce macroscopic fruiting bodies

Mushroom- Large, sometimes edible, fruiting body produced by some fungi

Mycology - the science and study of fungi.

Mycoplasma - A micro-organism intermediate in size between viruses and bacteria possessing many virus-like properties and not visible with a light microscope.

N

Negative staining - a staining procedure in which a dye is used to make the background dark while the specimen is unstained.

Nephelometer- Any apparatus used to measure the size and concentration of particles in a liquid by analysis of light transmitted through or reflected by the liquid.

Neutrophile – micro-organism living under neutral pH conditions (pH5.5-8.5).

Nitrogen fixation - the metabolic process in which atmospheric molecular nitrogen is reduced to ammonia; carried out by cyanobacteria, Rhizobium, and other nitrogen-fixing bacteria.

Nosocomial infection - an infection that develops within a hospital (or other type of clinical care facility) and is produced by an infectious organism acquired during the stay of the patient.

Nomenclature- System of naming organisms.

Non-conformity – a deficiency in a characteristic, product specification, process parameter, record, or procedure, that renders the quality of a product, raw material or process unacceptable.

Non-unidirectional air-flow – where the air supply entering a clean zone mixes with the internal air. See **turbulent air**.

Normal flora- The micro-organisms which ordinarily grow on the various surfaces of a plant or animal.

Nutrient- Substance taken by a cell from its environment and used in catabolic or anabolic reactions

Nutrient medium - a liquid broth or semi-solid jelly containing nutrients which stimulate and sustain the culture and proliferation of bacteria, higher plant cells or animal tissue.

Nutritive properties - the term given to the fertility testing / growth promotion of culture media in order to demonstrate that the test medium supports microbial growth. Can include general or selective testing.

O

Objectionable microorganism - see indicator microorganism

Obligate- (i) Adjective referring to an environmental factor (for example, oxygen) that is always required for growth. (ii) Organism that can grow and reproduce only by obtaining carbon and other nutrients from a living host, such as obligate symbiont.

Obligate aerobes - microorganisms that grow only in the presence of oxygen.

Obligate anaerobes - microorganisms that cannot tolerate the presence of oxygen and die when exposed to it.

Oligotroph- Microorganism specifically adapted to grow under low nutrient supply. Thought to subsist on the more resistant organic matter and be little affected by the addition of fresh organic materials. Sometimes a synonym for autochthonous.

Oligotrophic environment - an environment containing low levels of nutrients, particularly nutrients that support microbial growth.

OOS – out-of-specification result. A result outside the range of an approved specification. An OOS must be investigated to determine whether it is due to laboratory error, process error, or operator error. A judgment is then made to determine if the result is valid or invalid.

OOT – out of tolerance, in relating to calibration; out –of –trend, in relating to unexpected results from historical or statistical trends.

OQ – operational qualification. Documented verification that all aspects of a facility, utility or equipment operate as intended through all anticipated ranges.

Organism – a single, autonomous living thing e.g. micro-organism.

Osmotic Pressure - Pressure generated by the osmotic flow of water through a membrane into a (aqueous) phase containing a solute in a higher concentration.

P

Parenteral Drug (LVP, SVP) - A parenteral drug is defined as one intended for injection through the skin or other external boundary tissue, rather than through the alimentary canal, so that active substances they contain are administered, using gravity or force, directly into a blood vessel, organ, tissue, or lesion. They are infused when administered intravenously (IV), or injected when administered intramuscularly (IM), or subcutaneously into the human body.

Particle - an object of solid or liquid composition, or both, and of a certain size. It is a small piece of matter with a defined physical boundary. The size and number of particles are critical in order to assess a cleanroom.

Particle concentration - The number of individual particles per unit volume of air.

Particle counter - a device which measures the number of airborne particles of a given size by passing a sample of air through a focused light source. The presence of a particle causes dispersion of the light and the presence of a particular size can be detected using pre-defined criteria. At BPL two sizes of particles are measures: 0.5μm (a close approximation of the size of a bacterium) and 5.0μm (a close approximation of the size of a human skin cell).

Pasteurization - the process of heating milk and other liquids to destroy microorganisms that can cause spoilage or disease. Also, a general term for using mild heat to reduce microbial numbers in heat-sensitive materials.

Pass-through hatches - pass-through hatches protect critical environments while allowing transfer or materials to or from adjoining rooms. Typical installations are in the walls of cleanrooms and biological containment laboratories. In these applications, materials must be transferred with minimal loss of room pressure; and without the need for personnel movement between rooms.

Passage – when micro-organisms are cultured, each dilution step in vitro is one passage. Some tests will have a limit on the number of acceptable passages.

Pathogen - an organism able to inflict damage on the host it infects. Some microorganisms isolated are potentially pathogenic and all isolated in the Microbiology laboratory are treated as potentially hazardous.

Pathogenic - producing or capable of producing disease.

Penicillium - The genus of mold causing a zone of inhibition in an agar plate of bacteria. It is the organism, which produces natural penicillin.

Penicillins - a group of antibiotics containing a b-lactam ring, which are active against Gram-positive bacteria.

Peptidoglycan-Rigid layer of cell walls of bacteria, a thin sheet composed of N-acetylglucosamine, N-acetylmuramic acid, and a few amino acids. Also called murein. It is most prevalent in Gram-positive micro-organisms.

Peptones - water-soluble digests or hydrolysates of proteins that are used in the preparation of culture media.

Petri dish   - a shallow dish consisting of two round, overlapping halves that is used to grow microorganisms on solid culture medium; the top is larger than the bottom of the dish to prevent contamination of the culture.

pH - The pH value of an aqueous solution is a number describing its acidity or alkalinity. A pH is the negative logarithm (base 10) of the concentration of hydrogen ions (equivalent per liter). The pH value of a neutral solution is 7. An acidic solution has a pH less than 7, while a basic solution has a pH greater than 7, up to 14.

Phase-contrast microscope - a microscope that converts slight differences in refractive index and cell density into easily observed differences in light intensity.

Phenol coefficient test - a test to measure the effectiveness of disinfectants by comparing their activity against test bacteria with that of phenol.

Phenotype- Observable properties of an organism

Phylogeny- Ordering of species into higher taxa and the construction of evolutionary trees based on evolutionary (genetic) relationships

Plankton – free-floating micro-organisms e.g. bacterioplankton.

Plasmid - Self-replicating, extrachromosomal circular DNA molecules, distinct from the normal bacterial genome and nonessential for the cell survival under nonselective conditions. Some plasmids are capable of integrating into the host genome. A number of artificially constructed plasmids are used as cloning vectors.

Plasmids are present in bacteria or isolated from bacteria. *Escherichia coli*, the usual bacteria in molecular genetics experiments, has a large circular genome, but it will also replicate smaller circular DNAs as long as they have an "origin of replication"

Plate count- Number of colonies formed on a solid culture medium when uniformly inoculated with a known amount of sample, generally as a dilute soil suspension. The technique estimates the number of certain organisms present in a sample.

Polymerase chain reaction (PCR) - an in vitro technique used to synthesize large quantities of specific nucleotide sequences from small amounts of DNA. It employs oligonucleotide primers complementary to specific sequences in the target gene and special heat-stable DNA polymerases.

Population - an assemblage of organisms of the same type.

Pour plate- Method for performing a plate count of microorganisms. A known amount of a serial dilution or sample is placed in a sterile Petri dish and then a melted agar medium is added and the inoculum mixed well by gently swirling. After growth the number of colony forming units, on the surface or within the agar, can be counted

PQ – performance qualification. Documented verification of a facility, utility or equipment can perform as intended in meeting pre-determined acceptance criteria.

Precision - agreement of the results of several independent assays of the same sample. Usually reported as a standard deviation of a specified number of assays. Note that this is about how the measurements are related to each other, and does not indicate how near any one measurement or the mean of several measurements, is to the true value, if that is known.

Pre-filter - a filter positioned in front of another filter (e.g. a HEPA filter) to reduce the blockage of contaminants. It will have a lower removal rating than the main filter.

Preservative - A bacteriostatic or bactericidal agent added to some multiple dose parenterals and most cosmetics. Examples are benzalkonium chloride (BAC), formaldehyde, and thimerosol (merthiolate).

Prion - A protein molecule that lacks nucleic acid, that is, no DNA or RNA, often considered to be the cause of various infectious diseases of the nervous system (such as Creutzfeldt-Jakob disease and scrapie.) Very resilient, not easy to kill.

Prokaryote - a description for a kingdom of organisms where the members have a simple cell lacking a true nucleus, a nuclear envelope, and membrane enclosed organelles. The bacteria are members of this kingdom. The appendages of these cells provide mobility by flagella, attachment by pili and fimbriae and in some cells DNA transfer also by pili.. The cell envelope (outer covering) consists of three layers: the glycocalyx, the cell wall and the cell membrane. The composition of the cell wall determines the classification of the bacteria (gram positive, gram negative, lacking cell wall or chemically unique cell walls). Most bacteria have three general shapes: coccus (round), bacillus(rod), or spiral.

Pseudomonad- Member of the genus *Pseudomonas*, a large group of Gram-negative, obligately respiratory (never fermentative) bacteria.

Psychrophile - a microorganism that grows well at 0°C and has an optimum growth temperature of 15°C or lower and a temperature maximum around 20°C.

Psychrotroph - a microorganism that grows at 0°C, but has a growth optimum between 20 and 30°C, and a maximum of about 35°C.

Pure culture- population of microorganisms, which are identical and are composed of a single strain. Such cultures are obtained through selective laboratory procedures and are rarely found in a natural environment

Pyogenic- Pus-forming; causing abscesses

Pyrogen - fever is initiated when a circulating substance called pyrogen resets the hypothalmalic thermostat to a higher setting. Pyrogenic substances include bacterial endotoxins; extotoxins and viruses. Endotoxins are examined using the LAL test and all pyrogens can be examined using the rabbit pyrogen test.

Pyrogenic - Fever-inducing

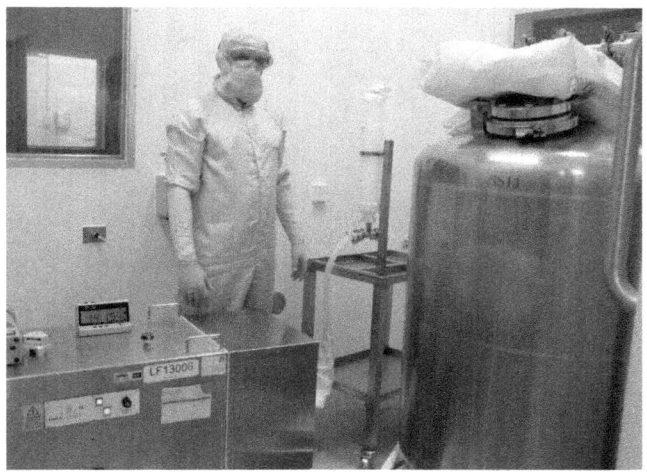

Q

Q10- A relationship for the effect of temperature on a process such that the process rate increases by the same multiple for every 10°C rise in temperature.

Qualification – Documenting that a piece of equipment does what it was designed to do. Was installed correctly, and continues to operate within specified parameters over time. The usual sequence of qualification steps is: DQ, IQ, OQ and PQ.

Quiescent - Not growing.

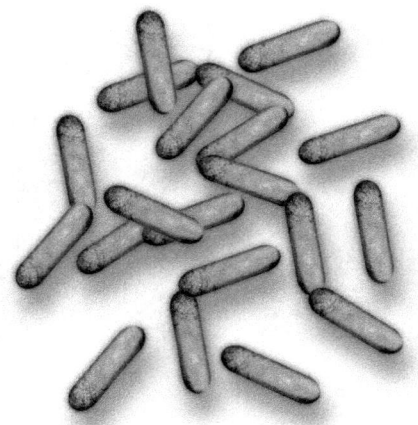

R

Radiation Sterilization - Sterilization using gamma radiation emitted from radioactive materials such as cobalt-60, or cesium 137. If proper dosage of nuclear radiation can be documented, sterility testing is not required.

Recirculating air - That portion of the workspace or cleanroom air that is recirculated through the air-conditioning equipment

Relative humidity - the ratio of the mole fraction of water vapor present in the air, to the mole fraction of water vapor present in saturated air at the same temperature and barometric pressure; approximately, it equals the ratio of the partial pressure or density of the water vapor in the air, to the saturation pressure or density, respectively, or water vapor at the same temperature.

Reference Standard, Primary - A substance that has been shown by an extensive set of analytical tests, to be authentic material of high purity. This standard may be obtained from a recognized source or may be prepared by independent synthesis or by further purification of existing production material.

Replication - the process in which an exact copy of parental DNA or RNA is made with the parental molecule serving as a template.

Reverse osmosis - The term reverse osmosis comes from the process of osmosis, the natural movement of solvent from an area of low solute concentration, through a membrane, to an area of high solute concentration if no external pressure is applied.

For reverse osmosis the process involves pushing a high solute solution through a filter that traps the solute, and many micro-organisms, on one side and allows the pure solvent to be obtained from the other side. This is by applying a pressure in excess of the osmotic pressure. The membrane – a thin film composite membrane - is semipermeable (it allows the passage of solvent but not of solute).

Ribotyping - Ribotyping is the use of *E. coli* rRNA to probe chromosomal DNA in Southern blots for typing bacterial strains. This method is based on the fact that rRNA genes are scattered throughout the chromosome of most bacteria and therefore polymorphic restriction endonuclease patterns result when chromosomes are digested and probed with rRNA.

Risk assessment – a systematic process of organizing information to support a risk decision. It involves the identification of hazards, analysis and evaluation of the risk associated with the exposure of the hazards. This can be quantified to the probability of occurrence of harm and the severity of that harm.

RODAC - is the international acronym for a contact plate (that is an agar plate, with a raised surface, applied to a surface for the enumeration of microorganisms). It stands for Replicate Organism Detection And Counting.

S

Sanitation - a cleansing technique used to remove microbes or other particles by reducing the level of contamination. It does not produce total kill. Involves the use of sanitization agents, which are chemicals/products such as soaps or detergents used as a cleaning technique to remove microorganisms and reduce the level of contaminants.

Saturated steam - steam holding all the moisture it can hold and still remain a vapor.

Scanning electron microscope (SEM) - an electron microscope that scans a beam of electrons over the surface of a specimen and forms an image of the surface from the electrons that are emitted by it.

Seed Lot - Seed Lot System - A seed lot system is a system according to which successive batches of a product are derived from the same master seed lot at a given passage level. For routine production, a working seed lot is prepared from the master seed lot. The final product is derived from the working seed lot and has not undergone more passages from the master seed lot than the vaccine shown in clinical studies to be satisfactory with respect to safety and efficacy. The origin and the passage history of the master seed lot and the working seed lot are recorded.

Selective medium- Medium that allows the growth of certain types of microorganisms in preference to others (by inhibiting the growth of unwanted microorganisms). For example, an antibiotic-containing medium allows the growth of only those microorganisms resistant to the antibiotic.

Separative devices - see Clean-air device, laminar flow . 'Separative devices' is the terminology adopted for ultra-clean enclosures in ISO 14644-7, Cleanrooms and associate controlled environments - Part 7: Separative devices (clean air hoods, gloveboxes, isolators and minienvironments).

Sepsis - The presence of various pus-forming and other pathogenic organisms or their toxins in the blood or tissues; septicemia.

serial dilution- Series of stepwise dilutions (in a diluent such as saline or a broth medium) performed to reduce the populations of microorganisms in a sample to manageable numbers.

Serotyping - a technique or serological procedure that is used to differentiate between strains (serovars or serotypes) of microorganisms that have differences in the antigenic composition of a structure or product.

Settle plate - an agar plate used for passive air-sampling. The plate is exposed for a fixed duration, after which it is incubated, and the numbers of microorganisms which have settled on it and have produced colonies are counted.

Slime - the viscous extracellular glycoproteins or glycolipids produced by staphylococci and Pseudomonas aeruginosa bacteria that allows them to adhere to smooth surfaces such as prosthetic medical devices and catheters. More generally, the term often refers to an easily removed, diffuse, unorganized layer of extracellular material that surrounds a bacterial cell.

Slime layer - a layer of diffuse, unorganized, easily removed material lying outside the bacterial cell wall.

Slime mold - a common term for members of the divisions Acrasiomycota and Myxomycota.

Soil-Unconsolidated mineral or organic matter on the surface that has been subjected to and influenced by genetic and environmental factors of: parent material, climate (including water and temperature effects), macroorganisms and microorganisms, and topography, all acting over a period of time and producing a product--soil--that differs from the material from which it is derived in many physical, chemical, biological, and morphological properties, and characteristics.

Southern blotting technique - the procedure used to isolate and identify DNA fragments from a complex mixture. The isolated, denatured fragments are transferred from an agarose electrophoretic gel to a nitrocellulose filter and identified by hybridization with probes.

Species- In microbiology, a collection of closely related strains sufficiently different from all other strains to be recognized as a distinct unit

Specific epithet- Designation of a particular organism in the binomial nomenclature system. For example, *coli* is the specific epithet of *Escherichia coli*

Specified microorganism - see indicator microorganism

Specificity – the degree to which a substance (e.g. disinfectant) exerts a definitive and distinctive influence on a particular process or micro-organism

Spirillum (plural, spirilli)- (i) Bacterium with a spiral shape which is relatively rigid. (ii) Bacterium in the genus *Spirillum*

Sporangiospore- Spore formed within a sporangium by fungi in the phylum Zygomycota.

Sporangium- Fungal structure which converts its cytoplasm into a variable number of sporangiospores; formed by fungi in the phylum Zygomycota.

Spores- Specialized reproductive cell. Asexual spores germinate without uniting with other cells, whereas sexual spores of opposite mating types unite to form a zygote before germination occurs (see also endospores).

Sporicidal - any agent with the ability to kill spores

Spread plate- Method for performing a plate count of microorganisms. A known amount of a serial dilution is spread over the surface of an agar plate. After growth the number of colony-forming units is counted

Stationary phase- Period during the growth cycle of a population in which growth rate equals the death rate.

Strain - a population of organisms that descends from a single organism or pure culture isolate.

Sterile - an absolute term meaning free of all living microorganisms but not necessarily any biologically reactive by-products such as exotoxins or endotoxins.

Sterility test - a test generally done on finished products which are to be injected.

Direct transfer: Samples of the product are placed in separate containers of an aerobic growth medium and an anaerobic growth medium and incubated for at least 14 days.

Membrane filtration: liquids (typically, the contents of 20containers) are forced, in a closed system, through two separate membrane filters (typically 0.45 μm pore size). Then aerobic and anaerobic growth media are placed on the separate filters and cultured for at least 14 days.

No growth is allowed in any of the media.

Sterilization- Rendering an object or substance free of viable microbes but not of microbial toxins, such as, endotoxin by destroying or removing all viable microorganisms, including viruses. An object cannot be slightly sterile or almost sterile- it is either sterile or not sterile. Control methods that sterilize are generally reserved for inanimate objects

Static state – refers to environmental or particle monitoring in a room which is unoccupied (but where equipment is operating normally). In the past this was called the 'unmanned state'.

Static contamination - occurs when a product contains non-metabolizing organisms. The numbers remain constant (<100/g) and normally represent a wide variety of species.

Sticky Mat - Floor mats which remove contaminates from shoes or wheels as they contact its sticky surface. Sticky mats come in a permanent and washable option or as peel-off tacky mats consisting of multiple layers of tacky plastic film which are peeled off as they get dirty. Available in multiple sizes and colors. Sometimes they are called 'tacky mats'. An alternative is fixed polymeric flooring.

Streak plate - a petri dish of solid culture medium with isolated microbial colonies growing on its surface, which has been prepared by spreading a microbial mixture over the agar surface, using an inoculating loop. BPL uses a modification of the Miles-Misra technique.

Succession- Gradual process brought about by the change in the number of individuals of each species of a community and by the establishment of new species that gradually replace the original inhabitants

Supernatant – material floating on the surface of a liquid mixture or the overlying fluid that remains after the precipitation of a solid component through centrifugation.

Surfactant – any substance that changes the nature of a surface, such as lowering the surface tension of water.

Swab - a wad of absorbent material usually wound around one end of a small stick and used for applying medication or for removing material from an area; also, a Dacron-tipped polystyrene applicator.

T

Taxon (plural, taxa)- A group into which related organisms are classified.

Taxonomy- Study of scientific classification and nomenclature.

Teichoic acids- All wall, membrane, or capsular polymers containing glycerophosphate or ribitol phosphate residues.

Terminal Sterilization - The process applied to product sealed in its final container that transforms a non-sterile product into a sterile one.

Tetracyclines - a family of antibiotics with a common four-ring structure, which are isolated from the genus Streptomyces or produced semisynthetically; all are related to chlortetracycline or oxytetracycline.

Thermal death time (TDT) - the shortest period of time needed to kill all the organisms in a microbial population at a specified temperature and under defined conditions.

thermophile- Organism whose optimum temperature for growth is between 45 and 85°C.

Thermophilic (Of A Microorganism) - With optimum temperature for growth above 45°C, many thermophilic bacteria exist at high temperatures (greater than 80°C) and many of their enzymes which posses high thermal stability, are of great commercial interest.

Total Organic Carbon (TOC) - A measure of the level of organic impurities in water by their carbon content that determines the operating life of activated carbon beds. This is one of the parameters used to determine the purity of Semiconductor Grade water. Feed water will have TOC measured in ppm (parts per million), and ultrapure water (UPW) will have TOC measured in ppb (parts per billion).

Total viable count - see viable count; the 'total' normally refers to the adding together of two or more tests to give the total number of micro-organisms present in a sample (for example, a fungal count on a selective agar like SDA to a bacterial count on a general agar like TSA); can be aerobic (TVAC) or anaerobic (TVAnC).

Toxin - a microbial product that can cause injury to other micro-organisms.

Toxoid - a bacterial exotoxin that has been modified so that it is no longer toxic but will still stimulate antitoxin formation when injected into a person or animal.

Transmissible spongiform encepalopathies (TSE) - neurological disease in mammals, caused by prions. Some culture media is required to be theoretically free of TSEs.

Turbulent air-flow - air that moves in a disorderly way and where the pathway is difficult to predict. This is often caused by air moving past an object which causes random eddies. This is opposite to **laminar** or **unidirectional airflow**.

U

UDAF – unidirectional air-flow. This indicates air moving in streamlines of the same direction but not necessarily at different velocities (for air moving in both the same direction and at the same velocity this is described as **laminar air-flow;** for air moving in different directions this is **non-unidirectional or turbulent air-flow**).

ULPA - ultra-low penetrating air-filter. A more advanced filter than a HEPA filter. It can reduce the number of particles in the air of 0.12 μm and greater by 99.999%.

Ultrafiltration - Molecular sieves; membranes with pores small enough to remove large molecules. Rated in terms of nominal molecular weight cutoff. A 10,000 Dalton (molecular weight) UF membrane, for example, will remove bacterial pyrogens that are typically in the range of 20,000 Daltons.

Ultramicrobacteria - bacteria that can exist normally in a miniaturized form or which are capable of miniaturization under low-nutrient conditions. They may be 0.2 mm or smaller in diameter.

Ultrapure Water - Water with a specific resistance higher than 1 megohm-cm. In the laboratory, it usually refers to Type I reagent grade water. Anything in laboratory water that is not H2O is an impurity. Although chemically pure water is not attainable, ultrapure water systems are now capable of reducing impurities down to the limits of detection.

Ultraviolet Radiation - Light in the wavelength region 200-300 nm, used to detect RNA or DNA that has the fluorescent dye, ethidium bromide, bound to it.

Utility Systems - Facility wide systems not tailored to a specific process and that do not have contact with the drug substance or potential drug substance.

V

Vector - An agent, such as an insect, that can carry a disease-producing organism from one host to another; the agent used to carry new genes into cells. Plasmids currently are the vectors of choice, though viruses and other bacteria may sometimes be used. These molecules become part of the cell protoplasm.

Vegetative cell- Growing or feeding form of a microbial cell, as opposed to a resting form such as a spore. The vegetative cell state is the form in which an organism is able to grow and divide continuously, given favorable conditions (although this can also refer to the non-reproductive phase in fungi). Unlike endospores, vegetative cells are relatively poor at surviving environmental stresses such as high temperature and drying

Viable- Alive; able to reproduce.

Viable but non-culturable- Organisms that are alive but cannot be cultured on laboratory media.

Viable count- Measurement of the concentration of live cells in a microbial population. There are four main methods: pour plate; spread plate; membrane filtration or most probable number. Often referred to as total viable count.

Vibrio- (i) Curved, rod-shaped bacterial cell. (ii) Bacterium of the genus *Vibrio*.

Virology - the branch of microbiology that is concerned with viruses and viral diseases.

virus - an infectious agent having a simple a cellular organization with a protein coat and a single type of nucleic acid, lacking independent metabolism, and reproducing only within living host cells.

Vricide – an agent that kills viruses.

VITEK - an automated microbiological identification system which detects reactions of microorganisms to a range of substrates.

W

Water - there are various types of water, categorized in part by their chemical and microbiological states:

- Water for Injection - is the highest grade of water. It can either be sterile or as bulk. There are limits on the number of viable microorganisms and endotoxin levels. It is produced by distillation at BPL.
- Demineralised water - is treated mains water where significant minerals and ions are removed. There are limits for the number of viable microorganisms.
- Mains (or raw or potable water) - is the incoming water to BPL from the water authority. There are limits for the number of, and absence of certain types of, viable microorganisms

Water activity (aw)- A measure of water content equivalent to percent humidity divided by 100.

Water Treatment - Water treatment, also referred to as water conditioning, can consist of adding or removing chemicals to change the properties of water. In water softening, for example, sodium ions are substituted for metallic ions that cause "hardness" thus reducing the scale-forming tendencies of water. Water purification on the other hand, always consists of removing undesirable impurities.

WFI - water for injection. Very pure water, produced by distillation and that meeting pharmacopoeial requirements for bioburden and endotoxin.

White Blood Cells - Spherical shaped cells which contain nuclei and comprise the smaller number of cells of the formed elements of whole blood. The major portion of the buffy coat is composed of white blood cells. White Cells are protective cells in the bloodstream. They attack bacteria by squeezing through capillary walls to reach the area of infection. (see Leukocytes)

Wild type-Strain of microorganism isolated from nature. The usual or native form of a gene or organism

X, Y, Z

Xenobiotic- A chemical which is not a natural component of the living organism exposed to it.

Xenobiotics - Industrial chemicals that have a chemical structure not found in natural compounds that may resist degradation by microorganisms.

Xerophile- An organism which can grow at very low moisture levels.

Yeast - a unicellular fungus that has a single nucleus and reproduces either asexually by budding or fission, or sexually through spore formation.

z value - the increase in temperature required to reduce the decimal reduction time to one-tenth of its initial value. See also decimal reduction time.

Zoonosis - a disease of animals that can be transmitted to humans.

Zygospore - A sexually produced resting fungal spore of the zygomycetes produced by the fusion of two morphologically similar gametangia.

www.ingramcontent.com/pod-product-compliance
Lightning Source LLC
Chambersburg PA
CBHW071628170526
45166CB00003B/1238